BIOLOGICAL EVOLUTIONISM

D1523798

SAINT DEMETRIOS CATHEDRAL
SUNDAY SCHOOL GRADUATION
ASTORIA, NY
June 2001

BIOLOGICAL EVOLUTIONISM

A lecture dealing with Lamarckian and Darwinian
Evolutionism that was delivered at Clark University in
Worcester, Massachusetts. Published in considerably
augmented form and amply documented.

By

CONSTANTINE CAVARNOS

INSTITUTE FOR BYZANTINE
AND MODERN GREEK STUDIES
115 Gilbert Road
Belmont, Massachusetts 02178-2200
U.S.A.

Design and Typesetting by:
E. Marshall Publishing & Translation Services
Brookline, Massachusetts 02146

First edition, 1994
Second, considerably enlarged edition, 1997
Published by THE INSTITUTE FOR BYZANTINE
AND MODERN GREEK STUDIES, INC.
115 Gilbert Road, Massachusetts 02178-2200, U.S.A.

Printed in the United States of America

Library of Congress Catalog Card Number: 97-71023
Clothbound ISBN 1-884729-24-X
Paperbound ISBN 1-884729-25-8

PREFACE

This book had its beginning in a course on "Problems of Philosophy" which I taught at Wheaton College in Norton, Massachusetts. The subject matter of the course consisted of problems such as free will vs. determinism, rationalism vs., empiricism, theism vs. atheism, creationism vs. evolutionism, and the soul-body relationship. I wrote out my lectures for that course, and when I later taught the same course at Clark University, in Worcester, Massachusetts, I considerably amplified my lecture on Biological Evolutionism.

In 1994, that lecture was published in booklet form at Etna, California. This second edition is augmented by the addition of considerable new material to the text, especially in the last two chapters, with more documentation, an Index of Proper Names, an Index of Subjects, and two beautiful illustrations showing Louis Agassiz and Alfred North Whitehead. For the photograph of Agassiz I owe thanks to the Harvard University

Archives. The drawing of Whitehead was taken from my book *Man and the Universe in American Philosophy* (*To Sympan kai ho Anthropos sten Amerikanike Philosophia*), which was published in 1959 at Athens by "Astir" Publishing Company and contains a chapter on Whitehead. The drawing was done especially for that publication by a Greek artist.

This book is addressed to all who are interested in the question of man's origin, nature, and destiny—philosophers and scientists, moralists and religionists, educators and students.

CONSTANTINE CAVARNOS

Belmont, Massachusetts
March 1997

CONTENTS

PREFACE v

1 INTRODUCTORY 9

2 LAMARCK'S THEORY OF
EVOLUTION 15

3 DARWIN'S THEORY OF
EVOLUTION 27

4 THE ORIGIN OF MAN 55

5 THE FRUITS OF BIOLOGICAL
EVOLUTIONISM 67

6 MAN'S DEGENERATION 75

INDEX OF PROPER NAMES 85

INDEX OF SUBJECTS 88

LOUIS AGASSIZ (1807-1873)
(Courtesy of the Harvard University Archives.)
Agassiz was one of the greatest and most popular scientists
of the 19th century. A naturalist, geologist, and educator,
"he made revolutionary contributions to the study of
natural science."

1

INTRODUCTORY

In this work I shall chiefly discuss the two dominant theories of biological evolution: Lamarck's and Darwin's. I shall explain what these theories assert and then subject them to critical examination. Before doing so, I shall say something about my qualifications to speak on biological evolutionism in a responsible manner.

I began my college education at Harvard as a pre-medical student. My courses during my first two years were chiefly in the natural sciences. Majoring in biology, I took courses in botany, zoology, comparative vertebrate anatomy, and physical anthropology. I also took courses in chemistry and a course in the history of modern science.

BIOLOGICAL EVOLUTIONISM

My mentors in these subjects were leaders in
their respective fields. One of them, Earnest Albert
Hooton (1887-1954), was the founder of the science
of physical anthropology and a leading propagator
of biological evolutionism, having written such
books as *Apes, Men, and Morons* and *Up from the
Ape*.

From the courses that I mentioned, I learned the
nature of the scientific method, both in theory and
in practice. For I both attended regularly the lectures
given by these teachers and also did a great deal of
laboratory work every week. In the biology courses,
I learned all about the evolutionism of Lamarck and
Darwin, and heard and read about speculations as to
the appearance of life and man on our planet. In the
history of science course, which was taught by
Giorgio de Santillana (1902—), I was shown in a
very clear manner the extremely important rôle
which seventeenth-century European philosophy
had played in the formation of various scientific
theories—astronomical, physical, and biological—
that appeared at that time and later.

10

Introductory

Towards the end of my sophomore year at Harvard, I changed my field of concentration to philosophy. I made this change because I was deeply troubled by some of the things that were taught in my biology courses, such as about the nature and origin of life and man, the mechanistic, materialistic conception of the living organism, including man, and Godless evolutionism. While still majoring in biology, I discovered and avidly read the writings of the German biologist and philosopher Hans Driesch (1867-1941), the Italian psychologist and philosopher Eugenio Rignano (1870-1930), and the French philosopher Henri Bergson (1859-1941). I found their critique of such views very telling and liberating. My tutor-advisor, Professor E.S. Castle, a physiologist, advised me not to read such works—he regarded them as scientific heresy—, but I did not heed his admonition. Dr. de Santillana's course in the history of modern science opened up for me vast intellectual horizons, those of general philosophic

11

thought, which was clearly something much broader and profounder than the sciences.

I earned my A.B., A.M., and Ph.D. degrees at Harvard in the field of philosophy. My philosophical training gave me a clearer, more precise understanding of the nature of the scientific method employed by the biological and other empirical sciences. It also awakened in me the perception of the *presuppositions* of these sciences. They are presuppositions of a philosophical, metaphysical nature and, as such, cannot be proved by the use of the scientific method. One of the presuppositions is that there is a stable, enduring order of nature, which the scientist can set out to study and describe. Without this presupposition, no natural science can exist or even get started.

My understanding of the evolutionism of Lamarck and Darwin, and of later followers of theirs—the Neo-Lamarckians and the Neo-Darwinians—, was improved considerably when I took a course on the philosophy of Henri Bergson. This course was taught by Ralph Barton Perry

(1876-1957), one of the leading twentieth-century American philosophers. (The course also included a study of the thought of the famous American philosopher William James, who was friendly to Bergson's ideas.) Early in this century, Bergson wrote a book, regarded as his most important one, entitled *Evolution Créatrice*. This was soon translated into English by a Harvard scholar. His version was published in 1911 in New York, under the title *Creative Evolution*. I read the entire volume of close to four hundred pages and listened carefully to what Professor Perry had to say about it. The fact that Bergson, a major professional philosopher, trenchantly criticized the evolutionary theories of Lamarck and Darwin and their followers, and presented an evolutionary theory of his own that commanded international attention, made it clear to me that the theories of biological evolution are *philosophical* in nature and should be treated as such.

During my career as a professor of philosophy, I have had occasion to discuss the Lamarckian and

BIOLOGICAL EVOLUTIONISM

Darwinian theories of evolution in my classes at Wheaton College (Mass.) and Clark University, in particular in my course "Problems of Philosophy." One of the most popular American textbooks for such courses has a division containing texts written by Lamarck and Darwin, and I was expected to assign these to my students and discuss them in class. The title of this book is *Philosophic Problems: An Introductory Book of Readings.* It was compiled and edited by Maurice Mandelbaum of Johns Hopkins University, Francis W. Gramlich of Darmouth College, and Alan Ross Anderson of Yale University. First published in New York by Macmillan Company in 1957, it has since been reprinted many times.

2

LAMARCK'S THEORY
OF EVOLUTION

The theory of biological evolution is usually associated with the name of the English naturalist or field biologist Charles Darwin (1809-1882). However, an ambitious attempt to develop a theory of evolution was made earlier by the eminent French botanist and zoologist Jean-Baptiste Lamarck (1744-1829). Lamarck was born sixty-five years before Darwin. In formulating his own theory, Darwin made use of some of Lamarck's ideas. The latter presented his evolutionary theory in his book *Zoological Philosophy (Philosophie Zoologique)*. This was first published in 1809—fifty years before Darwin's famous work *The Origin of Species.*

According to Lamarck, the more perfect or more complex species of plants and animals evolved from the less perfect or simpler species by a gradual, very slow process. The progression from the lower to the higher species was effected by various environmental influences and the reaction of the organisms to these influences. The principal circumstances involved were (a) the climate, (b) different temperatures, and (c) the diversity of the different localities.

These circumstances led animals to perform new actions, which became habitual. Thus, they developed new habits. Changes in the habits led to changes in the form and structure of organs and other bodily members.

According to Lamarck, these changes took place in accordance with *the law of use and disuse.* That is, an organ or part of the body is strengthened, developed, enlarged by use or exercise, whereas it weakens, atrophies, and disappears through disuse. These changes, which are changes in the form and structure, were preserved by *heredity* and passed on to the new individuals which descended from them.

Thus, gradually, new species of animals sprang or "evolved" out of the already existing species.

The sequence of causes involved in evolution is, briefly, as follows: New circumstances—new actions—new habits—new forms of organs (law of use and disuse involved)—perpetuation of these new organs (through heredity)—new species (as the differences of form become very pronounced).

An example of this process given by Lamarck are the shorebirds. According to him, the shorebirds acquired long legs because for many generations they made constant efforts to extend and elongate them, as they were in constant danger of sinking in the mud. Another example he gives are the giraffes. The giraffe, according to him, acquired its long neck and forelegs because its ancestors made continued efforts to reach the foliage of trees in regions destitute of herbs.

I mentioned the fact that Lamarck was a biologist. The question may be asked now: Is his view, thus outlined, *a scientific* one or is it one belonging to the sphere of *philosophical speculation*?

I submit that when Lamarck carefully observes and describes the anatomy of certain shorebirds, the giraffe, moles, and so on, he is doing science, biology. Similarly, he is doing science when he observes that use tends to strengthen and perfect organs, while disuse tends to cause their weakening and atrophy. This phenomenon has been observed by other scientists and by common men. However, Lamarck fails to express with sufficient scientific accuracy what observation discloses. Observation shows that use develops organs or bodily members, provided use does not go beyond a certain point. For it is also a fact that overuse does not strengthen but weakens, does not perfect but causes deterioration. Thus, excessive use of the stomach by overeating does not strengthen the stomach but weakens it, causes various stomach disorders.

Nor does use or overuse necessarily enlarge an organ: thus, the eyes of an adult do not become larger by frequent use, nor does his brain become larger as a result of great physical or intellectual activity. If they did, the sockets of the eyes would not be able to accommodate the eyes, and the skulls would crack from the pressure of the enlarged brains!

It is obvious that Lamarck did not formulate his principle of use and disuse with due accuracy. He spoke of use and disuse, but did not take cognizance of overuse and abuse, and of the fact that use does not necessarily result in enlargement or in structural changes.

Further, I submit that when Lamarck goes *beyond scientific observation* to the formulation of a theory of evolution, of certain species evolving from others—the higher from the lower—he is speaking as one who engages in *speculative thinking*, in *conjecture* and *imagination*. Thus, his explanations of how the shorebirds and the giraffes came to be what they are sound rather fanciful and unconvincing. They are obviously *purely hypothetical*. Neither he nor anyone else has actually observed such things happening.

19

I must add a word of explanation about "acquired characters" or "acquired characteristics," giving examples other than the shorebirds and giraffes, whose alleged (by Lamarck) changes no one has observed. By an "acquired characteristic" Lamarck means a change in organs or parts of the body developed in the course of an individual's lifetime in the "somatic cells"—the cells that constitute the body.

A well-known example is the enlargement of the blacksmith's arm muscles as a result of his continually using them as he pounds iron with a hammer. According to Lamarck's evolutionary theory, the children of the blacksmith should have larger, more muscular arms than other children. Another example is that of a person who exposes his body to the sun and acquires a deep tan year after year. His children should, on Lamarck's theory, be born with darker skins than other children of white parents who do not regularly acquire a tan. But there is no evidence at all to show either that the blacksmith's sons or daughters are born with better developed arms, or that the children of those who persistently tan their skin are born with darker skins.

Lamarck's theory depends, among other things, on the *assumption* that acquired characteristics—such as those just mentioned—are *transmitted* by parents to their offspring through *heredity*, and pass on again through *heredity* to succeeding generations. But this assumption has been shown to be untenable by the German biologist August Weismann (1834-1914). In his famous book, *The Theory of the Continuity of the Germ Plasm*, which appeared in 1885, Weismann held that the *germinal* cells—the ova and the spermatozoa—are independent of the *somatic* cells, those of the rest of the body. This means, if true, that the transmission of acquired characteristics—which transmission Lamarck assumed and asserted, in order to explain the origin of different species of plants and animals—is a *fiction*. Experiments in genetics since Weismann's time have again and again disproved the view that acquired characteristics are inherited.

21

Clearly, Lamarck's evolutionary theory is open to serious criticism on *scientific* grounds. But it is open to serious criticism on *philosophical* grounds as well, being—taken as whole—a piece of *philosophical speculation*. That it is such is evidenced by the very title of the book in which it is presented: *Philosophical Zoology*.

Viewed as philosophical theory, Lamarck's evolutionism may be characterized as *atheistic* in substance, and as such subject to the criticisms to which atheism has been subjected through the ages by philosophers and theologians. Lamarck seeks to explain the origin of life and the existence of various species of living things on our planet without reference to a cosmic plan or design, without reference to a World-Mind or God. He attempts to explain the appearance of the *first forms* of life through *spontaneous generation*, and not as being a result of Divine creation. *He neither offered nor could offer any real evidence or proof whatsoever for this view.* Since his time, all supposed instances of spontaneous generation of living forms have been scientifically disproved. Among those who have experimentally disproved supposed instances of spontaneous generation of living organisms is the famous French chemist Louis Pasteur (1822-1895).

Assuming the spontaneous appearance of the first living organism, Lamarck procceded to explain the development of other living forms without any reference to God. He dismissed the creationist view of life which involves God as the *Creator* and *Designer*. It is as though Lamarck said: "Let us *imagine* or *assume* that God does not exist, and then let us see how the various kinds of living beings might have come into existence."

Now actually Lamarck was not a complete atheist. He asserted God's existence and attributed to Him the creation of the *inanimate* world—the world that is devoid of life—but there his assertion about God stopped. God remained for Lamarck remote from the realm of living things, including man, neither bringing about their existence nor giving direction to the course of their development.

23

Lamarck's God is the God of "deism"—a *philosophic* doctrine fashionable in his time in Western Europe. The God of "deism" is a mere *abstract idea*, a mere *postulate*, useful only for those who want to develop a purely mechanistic cosmology. Such thinkers feel the need of a God Who merely creates *inanimate matter* and lets it develop according to purely mechanical principles.

Obviously, a person who believes in the real God, the "God of Abraham, of Isaac, and of Jacob," Who revealed Himself through the Prophets and Christ, cannot view God as irrelevant to any attempt to explain the first appearance of life on the earth and the nature and destiny of living things, especially of man.

Lamarck's evolutionary theory was adopted by the French zoologist Geoffroy Saint-Hilaire (1772-1844). But both Lamarck and Saint-Hilaire were vigorously criticized by the French naturalist Georges Cuvier (1769-1832), founder of the science of comparative anatomy. Cuvier maintained that there is a *stability of the species* of living organisms, whose origin he attributed not to evolution, but to Divine creation.

The evolutionary theory of Lamarck found many adherents in America during the last century up to the mid-1890s, in the form known as Neo-Lamarckism. In Soviet Russia Neo-Lamarckism was the favored form of biological evolutionism until the late 1940s.[1] American Neo-Lamarckism was often referred to as "The American School of Evolution."

Neo-Lamarckism emphasizes the influence of the environment, the inheritance of acquired characteristics, and purposeful adaptations. In addition—going beyond Lamarck—it asserts that there is a "vital force" in organisms, which is involved in the evolutionary process, but it fails to explain this force. The reason why Neo-Lamarckism was favored in America and Soviet

---

[1] See George Daniels, ed., *Darwinism Comes to America*, Waltham (Mass.), Toronto, London, 1968, pp. 75-76.

Russia over Darwinism was that Neo-Lamarckism served as a prop for the American and Soviet preconception that social "progress" was to be achieved by changing the social environment through various manipulations. Thus, a *preconception*, not a scientific observation, was involved in the espousing of Neo-Lamarckism.

LAMARCK

3

DARWIN'S THEORY
OF EVOLUTION

Turning to Charles Darwin (1809-1882), we must note at the outset that he accepted Lamarck's basic thesis that the higher species of living organisms have "evolved" from the lower, and that acquired characteristics are inherited. However, he introduced two additional principles in order to explain *how* the supposed evolution takes place. Darwin presented his views in his book *The Origin of Species*, which, as we noted, was published fifty years after Lamarck's *Philosophical Zoology*—in 1859.

This Englishman had more imagination and resourcefulness than Lamarck. Also, he gave a little more scope to God's creativity than did the Frenchman. In his account of living things, Darwin does not ignore God altogether: he expresses the belief that God *created one or a few living forms*, and that it was from this or these that the rest "evolved."

Apparently in anticipation of a hostile reaction to his theory of evolution by the theists or creationists, wishing to soften such a reaction, Darwin remarks: "There is grandeur in this view of life, with its several powers having been originally breathed by the Creator into a few forms, or into one; and that, whilst this planet has gone cycling in accordance to the fixed law of gravity, from so simple a beginning endless forms most useful and most wonderful have been, and are being evolved."

Darwin's followers, however, known as "Darwinians" and "Neo-Darwinians," in general dispensed with such a God altogether. Now while Darwin brings in God at the beginning of his theory, in order to account for the *origin of life*, from there on he dispenses with God. He proceeds to explain the appearance of subsequent forms—plants and animals—in a *purely materialistic and mechanistic way*. The idea of a Design and of a Designing God

28

is dispensed with. According to Darwin, new species appear on earth one after another as a result of purely mechanical forces, requiring no creative intervention on the part of God to bring about their appearance. Man himself appears on the scene as a result of purely mechanical processes, not as a creature of God, not in accordance with a Divine plan.

It should be remarked that originally Darwin did not include man, the human species, in his theory of the origin of species. But later on, in his book *The Descent of Man*, which was published in 1871, he included man among the products of evolution.

In presenting his evolutionary theory, Darwin speaks at the outset very much like his predecessor Lamarck. He speaks of "new conditions" causing "variations" in organic beings, and says that variations generally continue, once they have begun, for many generations. But he does not speak of, or cite, new "species" as appearing in this manner, but only new "varieties." Varieties are groups of plants or animals that rank below a "species;" their members are members of the same species.

The question is posed: How are "varieties" ultimately converted into "species"? Darwin's answer is that this is effected through (a) the "Struggle for Existence" and (b) "Natural Selection" or the "Survival of the Fittest." The "Struggle for Existence" and "Natural Selection or the Survival of the Fittest" are introduced as *new* principles—they are additional to those set forth and utilized by Lamarck.

Darwin borrowed his first principle, that of the "Struggle for Existence," not from a biologist, but from an English *political economist*, Thomas Malthus (1766-1834).[1] Malthus held that more *human* beings are produced than can possibly survive with the existing means of sustenance, and hence there is a *struggle* for existence. Darwin *dogmatically* gave *maximum extension* to this idea—he extended it

---

[1] *An Essay on the Principle of Population*, London, 1798.

from *one* species, the human, to *all*! He said: "A struggle for existence inevitably follows from the high rate at which all organic beings tend to increase. There is no exception to the rule that every organic being naturally increases at so high a rate that if not destroyed, the earth would soon be covered by the progeny of a single pair."

Having stated and explained his principle of the "Struggle for Existence," Darwin proceeds to bring in his second explanatory principle, that of the "Survival of the Fittest," or "Natural Selection." He asserts that in the struggle for existence, which is inevitable, owing to the overproduction of progeny by the various species of plants and animals, the *fittest* survive. Since many more individuals are born than can possibly survive by the available means of sustenance, those individuals that have any advantage, however slight, over others, have the best chance of surviving and of procreating their kind. There is thus a preservation of favorable individual differences or variations, and an elimination or destruction of the injurious. "Natural

Selection," says Darwin, "rejects the slightest variations, if they are bad, and preserves and adds up all the good, working at the improvement of each organic being in relation to its conditions of life."

He notes, however, that *we see nothing* of these slow changes in progress, *until ages have elapsed.* For a great amount of modification to result in a species, a "variety" when once formed must again vary or present individual differences of the same favorable nature as before; and these must again be preserved, and so on onwards, step by step.

Darwin's mode here of seeking to explain the origin of species is obviously *hypothetical.* He himself refers to it as an "hypothesis." Thus, he remarks: "But whether it is true, we can judge only by seeing how far the *hypothesis* [my italics] accords with and explains the general phenomena of nature." If, as Darwin admits, *we see nothing* of the slow changes over vast lapses of time, that precede the appearance of a new species, this means that Darwin simply *assumed*—did not observe—both the occurrence of such changes, their preservation over ages, and the accumulation of such favorable differences in members of a species. No experimental proof for such a thing is possible, as the accumulation of favorable differences is said to take place over a period of ages.

The *hypothetical* nature of Darwin's evolutionism is also apparent in the illustrations he gives of how evolution works out—i.e., of how new species evolve (according to him) out of earlier, simpler species. Thus, he remarks: "In order to make it clear how, as I believe, natural selection acts, I must beg permission to give one or two *imaginary* [my italics] illustrations." One of his "imaginary" illustrations is that of the swiftest and slimmest wolf surviving during the season when food is scarce.

Darwin holds, like Lamarck, that useful variations are transmitted through *heredity*. Parents with variations that have survival value tend to produce offspring similarly characterized. Such variations, according to Darwin, are always small. He states that "Natural Selection acts only by the preservation and accumulation of small inherited modifications, each profitable to the preserved being." He rules out "any great and sudden modifications in their structure."

Continuing his account of how new species, according to his theory, spring from older ones, he asserts that through the accumulation of variations, new species eventually appear. "Small differences distinguishing varieties of the same species, steadily tend to increase, till they equal the greater differences between species of the same genus."

One final point in Darwin's account of the evolutionary process is of greatest importance for understanding his *motivation* in formulating such a theory. It is this: it *facilitates* the biologist in *classifying* various species by *assuming* a kind of *cause and effect relationship* between one species and another. Thus, he says: "If species had been independently created, no explanation would have been possible of this kind of *classification*." In other words, the theory that the various species of living things originated successively from one another serves as a *convenient assumption or fiction for classifying them.*

Having thus summarized Darwin's theory of biological evolution, I shall now undertake to point out some of the criticisms that have been advanced against it *by scientists* and *philosophers.*

In the first place, like Lamarck's theory, Darwin's theory is an *hypothesis,* and *not–* as one often hears said—a "scientific fact." Strictly speaking, it belongs to "philosophy of nature." Darwin, as I remarked earlier, himself speaks of his own evolutionary theory as an "hypothesis." Now it is of the very nature of an hypothesis to be *tentative, provisional, subject to revision or even abandonment,* when serious objections are raised against it. Darwin himself was quite aware of this. In his *Diary* he remarked: "I have steadily endeavored to keep my mind free so as to give up any hypothesis, however much beloved, as soon as facts are shown to oppose it."

Those who speak of the Darwinian evolutionism as a "fact" are evidently unaware of

Darwin's use of an "*hypothetical mode of reasoning*," some instances of which I have already noted.

Secondly, what we *actually know* is (a) that some species, such as the dinosaurs and the mastodons, *have become extinct*, and (b) that countless other species—animals and plants—*persist in existence. No one has observed the appearance of a new species.* Darwin himself admitted this when he said that it takes *ages* for a new species to be produced. Obviously, then, the development of one species of plant or animal out of another *cannot be established by direct observation.* Hence, the theory of evolution remains *an hypothesis*, for which its supporters can only marshal supposed indirect evidence.

Thirdly, in his theory Darwin tries to get new species from *varieties*, from new "breeds." *But no one has really proved that species originate from varieties.* Observation shows that *varieties always remain within the limits of the particular species to which they belong.* Not only this, but varieties are known to be unstable, to *tend to revert after a number of generations, to the original type.*

Fourthly, Darwin stresses the "Struggle for Existence" as a *universal* factor among living beings. This we noted in his *sweeping generalization* of the thesis advanced by Thomas Malthus regarding the *human* species. Darwin's popularizers, such as the English biologist Thomas H. Huxley (1825-1895), have stressed this struggle even more than their teacher. But some of the critics of Darwinism have called attention to the fact that there is also an *opposite principle* observed in operation: "*Mutual Aid*" among living beings. Mutual Aid is as much a *fact* as the struggle of one member of a species against another member of it, or against the members of another species.

The Russian prince Kropotkin (1842-1921) was one of those who observed and remarked upon this fact. In the region which he visited in Eastern Siberia and Northern Manchuria, Kropotkin was struck by the many instances which he observed of *mutual aid and mutual support amongst animals.*

"War of each against all is not *the* law of nature," says Kropotkin. "Mutual aid is as much a law of nature as mutual struggle."[2] The prince adds that "the animal species in which individual struggle has been reduced to its narrowest limits, and the practice of mutual aid has attained the greatest development, are invariably the most numerous, the most prosperous, and the most open to further progress."[3]

The eminent English surgeon and author Kenneth Walker presents certain observations that confirm those of Kropotkin. In his journeys in East Africa, Walker noted among other significant phenomena the co-existence of giraffes and elephants. These animals were often found in company and apparently for a very good reason: the elephants had enormous ears and excellent hearing, but poor eyesight, while the giraffes were like sentries posted

---

2 Peter Kropotkin, *Mutual Aid: A Factor in Evolution*, edited by Paul Avrich, London, 1972, p. 49.

3 *Ibid.*, p. 246.

on the watchtowers. When they combined their capacities it was almost impossible to get near them without being heard or seen.[4]

Fifthly, Darwin's second main explanatory principle, that of "Natural Selection, or the Survival of the Fittest," is also open to serious criticism. *It arbitrarily excludes the idea of purpose, of plan, of a designing Mind.* The term "Natural Selection" actually means *mechanical selection*—selection by purely material, mechanical forces, totally unrelated to God or a non-material force.

In this respect, Darwin's theory is quite akin to the evolutionary theory of Empedocles (*ca.* 490-430 B.C.), one of the pre-Socratic Greek philosophers. Empedocles put forth the theory—whether seriously or in jest is not known—that animals evolved as follows: First, there grew out of the earth various separate organs and bodily parts: heads, necks, arms, eyes, legs, and so on and so forth. Then these

---

[4] *Meaning and Purpose*, 1944, p. 45.

began to wander about. "Heads without necks grew up, and arms roamed about naked bereft of shoulders, and eyes wandered about alone lacking foreheads. These things then fell together in whatever way each chanced, and many others also in addition to these continually came into being.... Wherever all parts came about just what they would have been if they had come to be for a purpose, such things survived, being organized spontaneously in a fitting way; whereas those which grew otherwise perished and continue to perish."[5]

In Empedocles' theory, as in Darwin's, there is clearly the idea that "variations" arise *spontaneously, not by purpose or by the action of mind, and are similarly organized.* And in both men's theories there is the idea that the unfit are eliminated and the fit survive. This is Darwin's theory in a nutshell!

Aristotle, who presents Empedocles' theory of evolution, rejects it as one that could not possibly be true, inasmuch as it views natural processes as

---

5 Aristotle, *Physics*, II. 8.

purely mechanical, non-purposeful. Like two other great philosophers of Antiquity, Socrates and Plato, Aristotle held that Nature is pervaded by *purpose*, directed to the good, drawn to God. Darwin reverted to Empedocles' fanciful speculation according to which the processes in the realm of living things proceed without purpose and the guidance of mind, by purely mechanical principles and chance. As Jacques Barzun remarks: "The notion of a Deity or Providence, or Life Force having a tendency of its own, or even a single individual having a purpose of its own, was ruled out by Darwin."[6] This is an obvious example of *Darwinian dogmatism*.

Sixthly, Darwin gives no satisfactory explanation of *how variations originate*. He says that they arise *spontaneously*, which means by mere chance. Here again purpose is left out, a non-material factor is left out, God is left out.

It is obvious that Darwin's biological evolutionism ultimately depends on variations that are

---

[6] Barzun, *Darwin, Marx, Wagner*, 1942, p. 14.

assumed—not demonstrated—to arise by mere chance.[7]

The inadequacy of Darwin's explanation of the *origin* of the various species of plants and animals was pointed out long ago by the versatile English writer Samuel Butler (1835-1902). Butler, who criticized Darwin's evolutionism on biological, social, metaphysical, and religious grounds, pointed out that: "*The operation of Natural Selection* [one of Darwin's chief explanatory principles] *might conceivably aid us to understand which forms survived, but it could never tell us how these forms had come to be.*"[8] What, asks Butler, *caused* the variations which provided Natural Selection with something to work on?

Sometimes Darwin sought to explain the appearance of variations not through mere chance, but through Lamarck's idea of use and disuse. This

---

[7] Cf. Edmund Sinnott, *The Biology of the Spirit*, New York, 1955, pp. 161-162.

[8] Barzun, *op. cit.*, pp. 117-118.

was despite the fact that he had sought to discredit Lamarck's work. Moreover, as we have seen, the belief that characteristics acquired through use are transmitted by heredity has been disproved again and again by the science of genetics.

Supposing we take Darwin's explanation that variations arise spontaneously. The question may be raised: Can such purely accidental variations, together with Natural Selection, account for the supposed evolutionary development of living forms? The philosopher Henri Bergson, whom I mentioned earlier, answers, "No." Bergson points out, among other things, that the eye of the pectin— a common mollusk—possesses a retina, a cornea, and a lens of cellular structure much like our own. He asks: How is it credible that a precisely similar chain of 'accidents' has happened in two such widely divergent branches of the evolutionary tree as the vertebrates and mollusks, which are invertebrates? "How could the same small variations, incalculable in number, have occurred in the same order on two independent lines of evolution, if

43

they were purely accidental? And how could they have been preserved by selection and accumulated in both cases, in the same order, when each variation taken separately, was of no use?"[9]

Some Neo-Darwinians, notes Bergson, have tried to get around this difficulty by developing the theory of *Orthogenesis*. This theory postulates the existence in the protoplasm of every animal an innate independent tendency which helps to maintain its evolutionary *progress in a certain direction.*[10] But what is this except to *postulate* (assume) a *purposive* factor, and thus to abandon the Darwinian theory of evolution by purely mechanical processes? The majority of recent Darwinians are inclined to reject such a theory. And thus they are left with the difficulty which Bergson and others have pointed out.

A seventh criticism that may be raised against evolutionism, and has been raised, is that it seeks to infer from the *similarity of form* which certain

---

[9] *Creative Evolution*, New York, 1911, pp. 56-57.

[10] *Creative Evolution*, p. 58.

species of plants and animals exhibit, a *common descent or origin* of the species. For example, man and the gorilla resemble one another, therefore man and gorilla "evolved" from the same species of animal.

Darwin, as we noted, remarked on the value of the evolutionary theory for *classifying* the various species of plants and animals according to certain *assumed* "cause and-effect relationships." But the fact that such a theory is a *convenient assumption* for classifying living things does *not* prove that the various species of plants and animals are in point of *fact* related in cause-and-effect relationships, i.e., that they are descended from one another. Such a relationship is merely a construction of the imagination of evolutionists.

The Darwinian classification of plants and animals in assumed cause-and-effect relationships, basing itself on similarity of form observable in certain species of plants and animals, has been *rejected* by the great Louis Agassiz (1807-1873). This Swiss-born scientist, "with an international

45

scientific reputation, came to the United States in 1845 and soon became the most popular scientist in the country."[11] In 1848, he was appointed Professor of Natural History at Harvard University. Ten years later, he founded the Museum of Comparative Zoology at Harvard.

Professor Agassiz maintained that the Darwinian classifications of various species of plants and animals into series according to their supposed origin (inferred from their similarities) are *arbitrary and fanciful*. He pointed out, for instance, that he found fossils of *higher fishes* in the *oldest* geological strata. Agassiz attributed the similarities of form between some species and others to the fact that *God so conceived them*. Each species of plant or animal was, according to Agassiz, created after a thought of God. Their resemblances or fundamental unities are to be viewed as rooted in the association of ideas in the Divine Mind.

---

[11] G. Daniels, *Darwinism Comes to America*, p. 2.

He held that there were successive *creations*, and that "there is *no evidence of a direct descent of later from earlier species in the geological succession of animals....* The most incontestable result of modern paleontological research... is the fact, now beyond controversy, of the *simultaneous* appearance of particular types of *all* classes of invertebrate animals from the earliest development of life upon the surface of the globe. The history of this successive development shows conclusively the *impossibility* of referring the first inhabitants of the earth to a *small* number of branches, differentiated from one parent stock by the influence of the modifications of external conditions of existence."[12]

The view advanced by Louis Agassiz to *counter* Darwin's assumption that similarities between two species implies the derivation (or "evolution") of one from the other, or of both from the same species, is quite in line with the creationist theory

---

[12] C.F. Holder, *Louis Agassiz: His Life and Work*, New York and London, 1893, pp. 182-183. [The italics are mine.]

presented by Plato and the creationism taught by Byzantine theologians, such as St. Dionysios the Areopagite.

In his *Timaeus*,[13] Plato taught that God fashions or makes the world out of matter contemplating an eternal, timeless "Model" (*Paradeigma*) that includes all the archetypes, "forms" or "ideas" of created things—in other words, of all the species of animate and inanimate creatures. Similarly, St. Dionysios the Areopagite says that the "forms" or "ideas," which he calls "models" (*paradeigmata*) of things, pre-exist in God, and that it was in accordance with these that God made and destined all creatures. Thus, he remarks: "We call *paradeígmata* the essence-bestowing pre-existing generative principles (*lógoi*) according to which the supra-essential Being predestined and made all creatures."[14]

---

[13] 48E ff.

[14] *Perí Theíon Onomáton* ("Concerning Divine Names"), Chapter V.

This view appears in other great theologians of the Eastern Orthodox Church, such as Maximos the Confessor (7th century), Symeon the New Theologian (11th century), Gregory Palamas (14th century), Markos of Ephesos (15th century), and Nicodemos the Hagiorite (18th century). The last mentioned Church Father says: "The distinctive forms inherent in creatures are effects and images of the uncreated archetypal forms which are in God."[15]

Another criticism—one of the most important —of the Darwinian theory of evolution is that it can give no satisfactory explanation of the presence of *mind in man*. Alfred Russel Wallace (1823-1913), a famous English naturalist, initially developed a theory of evolution similar to Darwin's at the same time that Darwin formulated his. But he abandoned it, when he was prompted to give due thought to this very serious defect.

---

[15] *Symbouleutikón Encheirídion* ("Handbook of Counsel"), Second edition, Athens, 1885, p. 165.

In his later period, Wallace held that man's *higher faculties* were not a product of evolution. Man's capacity for *abstract thought*, for *moral distinctions*, for *aesthetic appreciation*, his *aspiration towards moral ideals*—all these he asserted, *remain inexplicable in terms of the evolutionary theory.* The evolutionary theory, he said, may account for the appearance of the *human body, but not of the human mind.* It is necessary to explain the human mind or soul by reference to the Cosmic Mind or God.

In his book *The World of Life*, Wallace notes that in his *Contributions to the Theory of Natural Selection*, which was published in 1870, in the last chapter, "The Limits of Natural Selection as Applied to Man," he pointed out "that some of man's physical characters and many of his mental and moral faculties could not have been produced and developed to their actual perfection by the law of natural selection alone, *because they are not of survival value in the struggle for existence.*" This was a *rejection* of Darwin's conclusion that "even man's highest qualities and powers had been developed out of those of the lower animals by natural or sexual selection." It was a rejection which he says greatly "distressed" Darwin.[16]

Writing forty years after the publication of his *Contributions to the Theory of Natural Selection*, he proceeded to add to the above: "I maintain that some vast intelligence, some pervading spirit is required to guide the lower forces in accordance with a preordained system of evolution of the organic world.... Very important is the argument, set forth (in this book) in some detail, showing the absolute necessity of a creative and directive power and mind as exemplified in the wonderful phenomena of growth, organisation, and fundamentally of cell-structure and of life itself."[17]

---

[16] *The World of Life: A Manifestation of Creative Power, Directive Mind, and Ultimate Purpose*, new edition, London, 1914, p. 315.

[17] *Ibid.*, pp. 333, 339.

John Fiske (1842-1901), noted American philosopher and historian, adopted a position similar to that of Alfred Russel Wallace. In his book *The Destiny of Man* (1884), Fiske accepts Darwin's theory of evolution so far as subhuman species are concerned; but when he comes to man he abandons the theory. He remarks that consciousness "cannot possibly be the product of any cunning arrangement of material particles. The Platonic view of the soul, as a spiritual substance, an effluence from Godhood, which under certain conditions becomes incarnated in perishable forms of matter, is doubtless the view most consonant with the present state of our knowledge.[18] He expresses the same view in a later work, *Through Nature to God*.[19]

Edmund Sinnott, the eminent American botanist to whom I referred earlier, takes a similar stand. He accepts Darwin's theory so far as the

---

[18] *The Destiny of Man*, Boston and New York, 1884, pp. 42-43.

[19] Boston and New York, 1889, pp. 27-28.

plants and animals are concerned, but when he comes to man he draws the line. Sinnott attributes a *soul* to man—as do Wallace and Fiske. The soul for him is not a product of the material part of man, of the body, of the evolutionary process.[20] He expresses the view that "in the universe there is an organizing principle," and that "some of it dwells in each of us as his own soul—not a transient and temporary configuration of atoms and molecules and quanta, but part of an eternal, universal Spirit. The soul is the highest level of that goal-seeking, integrating process that is life. It is a magnificent hypothesis, and, as any good hypothesis should do, it accounts for many facts that otherwise would be quite meaningless.[21]

Obviously, Wallace, Fiske, and Sinnott considered *their view* regarding man as *superior, more in accord with the facts, than Darwin's hypothesis.*

---

[20] *The Biology of the Spirit*, New York, 1955, pp. 159-160.

[21] *Ibid.*, p. 160.

BIOLOGICAL EVOLUTIONISM

From what I have said thus far, it is evident that the theory of evolution as presented by Lamarck and Darwin and their followers belongs *not to science*, but to *free philosophical speculation*. It is not the product of any one science, for it claims to be based on facts of many sciences: *botany, zoology, morphology, comparative anatomy, genetics, embryology, geology, paleontology*. It cites facts which are *supposed* to confirm it, *ignores facts which have the opposite tendency*, and introduces *imaginary* "facts"—facts which no one has observed. It holds that new species of plants and animals originated from "varieties" within species, without citing a single *observed* instance of such a transition. I have given examples of each of these points.

DARWIN

4

THE ORIGIN
OF MAN

In connection with *man*, it should be noted that according to the evolutionary theory there should survive specimens of various *intermediate* forms, which intervened between man and his assumed sub-human ancestors. But all attempts to find such survivals, alive or dead (skeletons), have failed. The "link" or "links" between ape and man *have not been found anywhere* on our planet.

From time to time some skull or part of a skull, or some other bone, or a tooth, is found, which is triumphantly heralded by the evolutionists as a supposed remains of the hitherto *missing link*. But careful investigation shows that an error was made, or a fraud was committed.

These searches began over a hundred years ago. In Java, a skull was found in 1891 and was asserted to belong to the "pithecanthropos" (ape-man). But careful studies have shown that this skull really belonged to an ape, not to an "ape-man." Moreover, it was ascertained that the geological stratum where it was found was a stratum where full-fledged men were already in existence!

Earlier, in 1856, a skull was found in Neanderthal, Germany, that was hailed as the skull of the "missing link." But this skull was found to have a brain nearly three times as large as that of a gorilla and larger than that of an average European man!

In 1922, a skull was found in Peking, China, and some teeth also, as well as other skeletal remains. Later, many other remains were found there. Careful study showed that these belonged to genuine humans and not, as was supposed at the beginning, to ape-men.

Of course, in many books where evolutionism is discussed, one sees lined up not one but several "intermediate" forms between an authentic ape and an authentic man. These forms look progressively less and less apish and more and more humanlike.

But such intermediate forms or "links" are nothing but the creations of the *imagination* of the artist who drew them. Only ignorant people would take these as being depictions of actually discovered and observed creatures! The imagination of the artist who drew them conceived them in accordance with the forms *postulated* by the evolutionism of Lamarck and Darwin.

Their evolutionary theory seeks to explain human nature by studying man's *supposed non-human* ancestors. One of the first assignments in my course in physical anthropology at Harvard, given by Professor Earnest Albert Hooton, was to go to the zoo at Franklin Park, near Boston, observe the forms and behavior of the various types of apes there—gorillas, chimpanzees, orangutans, baboons, monkeys—and write a paper setting forth our observations.

The English poet Alexander Pope (1688-1744), possessing greater insight than the evolutionists, remarked in his *Essay on Man: "The proper study of mankind is Man."* This is a good lead for the anthropologist, the student of human nature. What needs to be added to Pope's statement is that the "Man" who should be studied is *the individual who has developed his distinctive human capacities: the intellectual, the aesthetic, the moral, and the spiritual to a very high degree.* So studied, man is seen to be separated from the irrational animals by an abyss. These capacities have their seat in the *psyche*, the *soul, not the body, on which the evolutionists focus their attention.*

Adopting this approach in their study of man, the great ancient Greek philosophers defined "man" *not* in terms of his *bodily* form: e.g., "Man is a featherless biped," but in terms of man's *inner essence*, his *soul*, saying: "Man is a *rational* animal." "Reason" (*logos, nous*), which encompasses all the distinctive human capacities that I mentioned, sets man apart from all the other species of animals. This is the definition of man given by Socrates, Plato, and Aristotle.

They arrived at this classical definition of man *not* by studying the *apes* and *speculating* about "links" between the apes and man, but by studying the *inner man*, the *soul*, both in themselves and in others. Plato had before him over a period of many years an *exemplary man*: his teacher Socrates; and Aristotle had before him for many years a similar teacher, Plato. The Orthodox Christian regards *the great Saints* of the Church as exemplary men par excellence, persons who have shown through their life, character, and thought *the true nature of man.*

Plato viewed man as a *heavenly creature*. Thus, in the *Timaeus* he says: "We are not an earthly but a heavenly plant."[1] He did not say this as a mere poetic metaphor; for he asserts in the same dialogue that the human mind is a creature of the supreme God, the "Demiourgos."

This teaching of Plato is diametrically opposed to that of Lamarckism and Darwinism, according to which man is *wholly an earthly creature*, having "evolved" from an ape, from an animal devoid of

---

[1] 90A.

anything heavenly, through a process in which God played no part whatsoever.

Aristotle held essentially the same view of man as Plato. That which differentiates man from the beasts, asserts Aristotle, is *nous*, the mind. This is not of earthly origin. *Nous comes from another realm (thyrathen)* and enters the earthly body at conception.[2] He also remarks: "It may be held that mind is *the true self* of each human being, inasmuch as it is the dominant and better part."[3] "Mind, more than anything else *is man (anthropos)*."[4] "And this is either itself actually divine, or is relatively the divinest part of us."[5] *Nous*—the mind or rational faculty—is the seat of the intellectual virtues and the fount of the moral virtues.

Ignoring the highly enlightened view of such great philosophers, and seeking to explain human nature in terms of their preconception of man being

---

2 *On the Generation of Animals*, 737a5.

3 *Nichomachean Ethics*, Book X. 1178a2-3.

4 *Ibid.*, 1178a15-8.

5 *Ibid.* 1177a15-17.

a product of evolution from lower forms of life, the evolutionists give a debased view of man. Ignoring man's God-like mental capacities: *cognitive, deliberative, moral, religious, aesthetic, creative,* they endeavor to present man as a being that differs from the irrational animals only in degree, not in his essential nature. They regard man as nothing more than an improved ape!

The need of studying man's *inner essence*, if one is to study man properly, has been emphasized by the eminent philosopher and theologian Nectarios, Metropolitan of Pentapolis, better known as St. Nectarios of Aegina (1846-1920). In 1893, twenty-two years after the appearance of Darwin's work *The Descent of man*, Nectarios published a book entitled *Sketch Concerning Man (Hypotyposis perí Anthrópou)*. In the Introduction he remarks:

"We must be profound inquirers and careful examiners.... The one-sided inquiry of the *body* [seen in the evolutionist theory] can lead to very imperfect and erroneous conclusions,which differ little from superficial observation. In order to be known, man must be examined *in depth*, not on the surface. We must come to know *his spiritual powers, his spiritual life*, his way of life, his relations to the universe and to his Creator. We must examine the purpose of his appearance in the world, his destiny. In this manner we shall be able to know him according to his dignity, in this manner we shall be able to make declarations about him.

"Those who have done this work with exactitude declare that man is a *rational being* that has a great destiny on the earth, that he has a heavenly soul, and that only as far as his body is concerned does he belong to the earth.

"The validity of the views of those who think in this way has the weight that is demanded by true reason and true science. This truth we wish to prove by studying man first *externally* and then *internally*."[6]

---

[6] *Op. cit.*, second edition, Athens, *ca.* 1978, pp. 9-10.

The Origin of Man

In the chapters that follow, Nectarios proceeds to examine man in the holoscopic (all-encompassing) way just indicated. His examination of man culminates in a discussion of the nature and genesis of the human soul, and in eleven proofs of its immortality, a subject about which both Lamarck and Darwin keep silent.[7]

He makes reference to Lamarck's evolutionary theory as presented in the French edition *Philosophie Zoologique*. He also mentions Darwin's evolutionism, about which he learned by reading Darwin's book *The Descent of Man* and a book on Darwinism written by the Greek author S.P. Sounkras.[8]

Criticizing Lamarck's theory, St. Nectarios says among other things the following: "The two volumes of the work *Philosophie Zoologique* are in their entirety intended to uphold the degrading

---

[7] These proofs appear in English translation in my book *Modern Greek Philosophers on the Human Soul*, Belmont, MA, 1967, pp. 59-86.

[8] *Sketch Concerning Man*, pp, 118, 166-167.

evolutionary theory regarding man. The first volume seeks to prove that the human organism evolved from that of the ape, as a result of chance circumstances. And the second volume seeks to prove that the distinctive excellences of the human mind are nothing but an extension of a power which the animals also have, differing only in degree. Having weak and badly set foundations..., Lamarck claims to prove that in earlier times nature produced through marvelous evolution one species from another, earlier one. He seeks to establish a gradual chain having successive (not contemporaneous) links and thus to produce finally the human species through a metamorphosis that is the reverse of the truth, and not less marvelous than the transformations one reads about in myths!"[9]

No details of the Darwinian hypothesis of evolution are discussed by Nectarios. He simply mentions the fact that like Lamarck, Darwin viewed man as a "perfected ape"! He makes the statement

---

[9] *Ibid.*, pp. 87-88.

that "the Darwinian theories imagined that they arrived at the solution of the anthropological question by accepting the mode of evolution. These theories, not being based on sound foundations, instead of solving the problem rendered it more enigmatic; because they denied the validity of revealed truth, viewed man as belonging to the same order as the irrational animals, denied his *spiritual origin* and attributed to him a very lowly origin. Their failure had as its chief reason the negation of his lofty origin and of his *spiritual nature*, which is altogether alien to matter and to the physical world. In general, without the acceptance of revealed truth, man will remain an insoluble problem. The acceptance of it is the firm and safe foundation upon which every inquirer about man must base himself. It is from this that he must begin in order to rightly solve the various parts of the question and learn the truth by means of true science."[10]

As we have seen, Plato and Aristotle say something close to the revealed truth regarding the

---

[10] *Ibid.*, p. 88.

nature of man. Both these philosophers assert, as we noted earlier, that the human soul is not of earthly origin, and has something God-like about it. This is akin to the teaching in the Book of Genesis, where it is said that man was created by God in His image and likeness. Nectarios attributed this kinship of their teaching to the fact that these philosophers were at times divinely illuminated, because they were lovers of truth and defenders of it.[11]

---

[11] Cf. St. Seraphim of Sarov: "The presence of the Spirit of God also acted among pagans who did not know the true God, although it did so less strongly than among God's people. Indeed, even among them [the pagans] God found for Himself chosen people. Such for instance, were the pagan philosophers who—although they wandered in the darkness of ignorance of the Deity—sought the truth, which is beloved of God. By this very God-pleasing search they were enabled to partake of the Spirit of God" (Constantine Cavarnos and Mary-Barbara Zeldin, *St. Seraphim of Sarov*, Belmont, MA, 1980, pp. 104-105).

5

THE FRUITS OF
BIOLOGICAL EVOLUTIONISM

By Divine source we are taught to judge teachings *"by their fruits."* Christ says: "Beware of false prophets, which come to you in sheep's clothing, but inwardly they are ravening wolves. Ye shall know them by their fruits.... Every good tree bringeth forth good fruit; but a corrupt tree bringeth forth evil fruit."[1]

Examining the *fruits* of the most widely accepted theory of biological evolution, Darwinism, we see that they have been *evil*. Even some Darwinians, such as Earnest Albert Hooton, explicitly admit this. Thus, in his book *Apes, Men, and Morons*, Hooton says: "I have been inclined to think that, on the whole, the dissemination of evolutionary

---

[1] Matthew 7: 15-17.

teaching...to the public at large may be inexpedient. An inept presentation of evolution to persons of limited mentality [i.e., the public at large] is likely to destroy their religious beliefs and fears, and to free them of inhibitions which make them socially tolerable."[2]

Regarding religion, which he says evolutionism tends to destroy in "the public at large," Hooton remarks: "I am convinced that religion...is a most efficacious and probably indispensable instrument for the shaping of human society."[3]

A fruit of the destruction of religious faith by evolutionist teaching is, according to Hooton, *antisocial behavior*, i.e., *moral degeneration*. The great Russian philosopher and writer Feodor Dostoievsky, expressed this consequence of such teachings in these few words: "If God does not exist, everything is permitted." That is, if one believes that the

---

[2] New York, 1937, p. 7.

[3] *Ibid.*, p. 6.

real, Biblical God does not exist, and believes instead in the imaginary God of the Deists, such as Lamarck and Darwin, a deity that neither created man nor judges him, the imperatives of morality are without binding force.

Hooton further admits that there have been, and there are, "mischievous propagandists who have attempted to use the evolutionary hypothesis as a weapon wherewith to *attack and destroy* established systems of *religion* and *ethics*."[4]

The abandonment of religion and its associated ethics, according to Hooton's own admission, has as a consequence the *moral degeneration of man*. The *fruits* of this degeneration are, he says: "The breakdown of free institutions, the disruption of decent human relations, the inadequacy of economic systems."[5] The "social doctors," he remarks, are fussing with such *symptoms*, and do not look to the *cause* of these symptoms. "The religionists are clos-

---

[4] *Apes, Men, and Morons*, p. 6. [My italics.]

[5] *Ibid.*, p. 3.

est to the heart of the matter, because they alone are seriously concerned with human ethics."[6] Who are the "*mischievous propagandists* that have attempted to use the evolutionary hypothesis as a weapon wherewith to attack and destroy established systems of religion and ethics?" Hooton does not cite any names. But an eminent contemporary Greek theologian, Nikolaos Pan. Vasileiadis, calls attention to a good number of such persons in his book *Darwin and the Theory of Evolution*.[7] He mentions among others Thomas Huxley, Marx, and Nietzsche.

Huxley (1825-1895) was the first who contributed decisively to the dissemination of Darwin's theory. Being a materialist and having observed that Darwin's ideas served as a prop to his atheistic, materialistic philosophy as well as to his struggle against the clergy, he became the most fanatical devotee of Darwin.

<hr>

[6] *Ibid.*, pp. 3-4.

[7] *Ho Darbinos kai he Theoria tes Exelixeos*, Athens, 1985.

The Fruits of Biological Evolutionism

Marx (1818-1883), the leading theoretician of Communism and fanatical enemy of religion, espoused Darwin's evolutionism with the greatest enthusiasm. The reason for this was the fact that Darwin's principles of the "Struggle for Existence" and the "Survival of the Fittest" served to support Marx's atheism and his ideas of "class struggle" that would lead to the victory of the "proletarian class," the laborers. For this class is the "fittest," i.e., the strongest, by virtue of its overwhelming numerical superiority over the rich. The French existentialist philosopher and literary writer Albert Camus (1913-1960) wrote in one of his works: "Marx wrote to Engels that the theory of Darwin constituted the foundation of their [Communist] theory.[8]

Nietzsche (1844-1900), the anti-Christian German philosopher who absurdly proclaimed "the death of God," was deeply influenced by Darwin and espoused his evolutionism. The related thought

---

[8] Vasileiadis, *op. cit.*, p. 120.

of Nietzsche played an important role in the formation of the Nazi worldview.

Vasileiadis remarks that "the mind and the actions of Hitler and all the Nazis were imbued with the Darwinism. ...Hitler's book *Mein Kampf* ('My Struggle') is based on Darwin's theory. In 1933, Hitler cried at Nuremberg that a *superior race* will always *conquer an inferior race*; and that the right of the stronger must prevail always and everywhere. He exterminated the Jews basing himself on this principle, because he considered the German race superior to all the others and destined to dominate everywhere and over all."[9]

Mussolini's Fascism, too, was inspired by Darwin's teaching of the "Struggle for Existence" and the "Survival of the Fittest." He often referred to Darwin in his speeches.

In his profound book *Science and the Modern World*, the great Anglo-American philosopher Al-

---

[9] *Ibid.*, p. 123.

The Fruits of Biological Evolutionism

fred North Whitehead makes certain important remarks regarding some of the *disastrous* consequences of nineteenth century *materialistic science and philosophy*—of which Biological Evolutionism is a very noteworthy example. He says: "In regard to the *aesthetic needs* of civilised society the reactions of science have so far been unfortunate. Its *materialistic basis* has directed attention to *things* as opposed to *values*. The antithesis is a false one, if taken in a concrete sense.... Thus all thought concerned with social organisation expressed itself in terms of material things and of capital. *Ultimate values were excluded....* A creed of competitive business morality was evolved...entirely devoid of consideration for the value of human life."[10] The scientific creed was that "matter in motion is the one concrete reality in nature."[11] "Ultimate values"—aesthetic, ethical, religious—were regarded as form-

---

10 *Science and the Modern World*, New York, 1948 (first edition 1925), pp. 291-292. [The italics are mine.]

11 *Ibid.*, p. 293.

ing "an adventitious, irrelevant addition."[12]

Then Whitehead adds that during the last three generations—those prior to 1925, when his book was first published—the exclusive direction of the attention of people to the aspect of "struggle for existence" has been "a disaster of the first magnitude. The watchwords of the nineteenth century have been, struggle for existence, competition, class warfare, commercial antagonism between nations, military warfare."[13]

These are some of the "evil fruits" that have been brought forth by the "corrupt tree" of Biological Evolutionism. Others could be added.

---

[12] *Ibid.*

[13] *Ibid.*, p. 295.

6

MAN'S DEGENERATION

The subject of man's *degeneration* calls for further discussion, particularly with reference to the failure of Darwin's hypothesis to recognize the reality or possibility of the *inverse of evolution.*

"Evolution" is defined as "a process of continuous change from a lower, simpler, or worse, to a higher, more complex, or better state" (Webster's Dictionary). This is the meaning "evolution" has for Lamarck and Darwin. Thus, Darwin says:

"Natural Selection is daily, hourly scrutinizing throughout the world, every variation, even the slightest; rejecting that which is bad, preserving and adding up all that is good; silently and insensibly working, whenever and wherever opportunity offers, at the improvement of each organic being, in its organic and inorganic conditions of life."[1]

This *dogmatic* assertion is *contradicted* by the *growing degeneration* of man—physical, mental, and moral. Such degeneration is testified to by reputable, scientific writers, such as Alexis Carrel, Weston A. Price, and Earnest Albert Hooton. I shall present some of their characteristic observations regarding modern man's degeneration.

Carrel (1873-1944), world-renowned biologist and surgeon, speaks again and again of human degeneration in recent times in his famous book *Man, the Unknown.* This book was first published in 1935 in New York and London, and has been reprinted innumerable times since then. He speaks of *physical, mental,* and *moral degeneration*—"degeneration of body and soul."

About physical degeneration Carrel particularly calls attention to the increase of "degenerative diseases, such as *cancer, heart lesions*, and *nephri-*

---

[1] Quoted in George Daniels, *Darwinism Come to America*, p. 36.

tis (diseases of the kidneys)."[2] Speaking of mental and moral degeneration, he observes: "Modern life has *decreased the intelligence and the morality* of the whole nation.... We pay several billions of dollars each year to fight criminals. Gangsters continue victoriously to attack banks, kill policemen, kidnap, ransom, assassinate children.... There are so many feebleminded and insane among civilized people."[3] "In practically every country there is a decrease in the intellectual and moral caliber of those who carry the responsibility of public affairs."[4]

Weston A. Price wrote a book, entitled *Nutrition and Physical Degeneration*, which was published four years after Carrel's *Man, the Unknown*. Price was an eminent dentist, a Member of the Research Commission of the American Asso-

---

[2] *Man, the Unknown*, 1935, p. xi.

[3] *Ibid.*, p. 275. [The italics are mine.]

[4] *Ibid.*, p. 21.

ciation of Physical Anthropologists. As a research scientist, he traveled to many parts of the world in order to discover the causes of dental caries and the degeneration of the jaws and other parts of the face. He had read Carrel's book and expressed agreement with the theses of Carrel which I have mentioned. Also, he was in touch with Hooton. The latter contributed a Foreword, in which he remarks that Price was a "really gifted scientist" and his book "profoundly significant."[5] In the course in Physical Anthropology which I took at Harvard, Price's book was assigned by Hooton as part of the required reading.

In the first chapter, Weston Price cites as examples of "certain tragic expressions of our modern degeneration tooth decay, general physical degeneration, and facial and dental-arch deformities."[6] And further on, he says: "Our American hu-

---

[5] *Nutrition and Physical Degeneration*, New York and London, 1939, pp. vii-viii.

[6] P. 5.

man stock has declined rapidly within a few centuries, and in some localities within a few decades.... No era in the long journey of mankind reveals in the skeletal remains such a terrible degeneration of teeth and bones as this brief modern period records."[7] "That this problem of serious degeneration of our modern civilization," he goes on to explain in the second chapter, "is not limited to the people of the United States has been commented on at length by workers in many countries."[8]

Although the book focuses on the problem of *physical* degeneration, its causes, and means of stopping it, Price also calls attention, here and there, to the seriousness of *mental* and *moral* degeneration.[9] And he devotes in the latter part of the book a whole chapter to a discussion of all three of these forms of degeneration.[10]

---

[7] P. 7.

[8] P. 11.

[9] P. 14.

[10] Pp. 353-381.

Hooton deals extensively with human degeneration in his already mentioned book *Apes, Men, and Morons*. At the beginning of the preceding chapter I called attention to his admission that the dissemination of evolutionary teaching to the public at large has had as a consequence the *moral degeneration of man*, whose fruits are "the break-down of free institutions, the disruption of decent human relations, and the inadequacy of economic systems." As a physical anthropologist, he dwells on the *physical* aspects of man's degeneration. And as these are related to man's *mental* degeneration, he occasionally calls attention to the latter, too.

With regard to the physical degeneration of man, Hooton remarks that an examination of the various parts of the human body "would *not show a single improvement* in man's anatomical status during the last 25,000 years, as revealed by the skeleton. *The brain has not increased in size (if anything, it has become smaller)....* Our reviewer of skeletal evolution would find certain clear evidence of *anatomical degeneration. The jaws have gone on shrinking* until often they are too small to accommodate the permanent teeth.... Frequently the lower jaw has failed to develop forward, so that the upper teeth protrude in a forlorn and rabbity fashion, while the chin retreats feebly into the neck.... *A horrible process of decay has infected the teeth,* beginning even in infancy.... *The nasal skeleton has continued to contract* laterally and to arch degeneratively upward, so that the septum becomes skewed to one side.... *The nose and the joints* of modern man are indubitably *out of joint.*"[11]

Hooton goes on to say that "with the development of urban civilization and industry, *physical deterioration becomes appallingly common.* All sorts of pathologies multiply." And he adds: "It is not commonly known, however, that *intelligence declines* with organic deterioration."[12]

---

[11] *Apes, Men, and Morons*, pp. 291-292. [The italics are mine.]

[12] *Ibid.*, p. 294. [The italics are mine.]

We may now go back to Darwin's highly optimistic, dogmatic statement, regarding the course of evolution which I quoted at the beginning of this chapter:

"Natural Selection is daily, hourly, scrutinizing throughout the world, every variation, even the slightest; rejecting that which is bad, preserving and adding up all that is good; silently and insensibly working, whenever and wherever opportunity offers, at the improvement of each organic being, in its organic and inorganic conditions of life."

In the light of the scientific data of Carrel, Price, and Hooton which I have presented, Darwin's statement is manifestly out of touch with reality, is pure optimistic fantasy. Together with other sweeping generalizations and assumptions of Darwin which I have pointed out in the course of this treatise, it provides an excellent example of what the Harvard philosopher A.N. Whitehead says regarding the use of the method of *philosophic speculation when used by modern natural science.* He makes the following observations:

ALFRED NORTH WHITEHEAD (1861-1947)
One of the greatest 20th century philosophers, Whitehead
undertook to develop a comprehensive philosophical
system, bringing into agreement the data of science with the
insights of philosophy (especially those of Plato),
the fine arts, ethics, and religion.

BIOLOGICAL EVOLUTIONISM

"In its use of this method natural science has shown a curious mixture of rationalism and irrationalism. Its prevalent tone of thought has been ardently rationalistic within its own borders, and dogmatically irrational beyond those borders. In practice such an attitude tends to become a dogmatic denial that there are any factors in the world not fully expressible in terms of its own primary notions.... Such a denial is the self-denial of thought."[13]

---

[13] *Process and Reality: An Essay in Cosmology*, New York, 1941, p. 8.

INDEX OF PROPER NAMES

Agassiz, Louis, viii, 45-48
Anderson, Alan Ross, 14
Aristotle, 40-41, 58, 59, 60, 65-66

Barzun, Jacques, 41
Bergson, Henri, 11, 12-13, 43-44
Butler, Samuel, 42

Camus, Albert, 71
Carrel, Alexis, 76-77, 82
Castle, E.S., 11
Cuvier, Georges, 24-25

Daniels, George, 25, 76
Darwin, Charles, 9, 10, 12, 13, 14, 15, 27-54, 57, 59, 63-
    65, 67-84
Dionysios the Areopagite, St., 48
Dostoievsky, 68-69
Driesch, Hans, 11

Empedocles, 39-40
Engels, 71

Fiske, John, 52, 53

Gramlich, Francis, W., 14
Gregory Palamas, St., 49

Hitler, Adolph, 72
Holder, C.F., 47
Hooton, Earnest Albert, 10, 57, 67-68, 76, 78, 80-83
Huxley, Thomas H., 37, 70

James, William, 13

Kropotkin, Peter, 37-38

Lamarck, Jean-Baptiste, 9, 10, 12, 13, 14, 15-26, 27, 29, 30, 33, 35, 42-43, 54, 57, 59, 63-65, 69, 75

Malthus, Thomas, 30-31, 37
Mandelbaum, Maurice, 14
Markos of Ephesos, St., 49
Marx, Karl, 70-71
Maximos the Confessor, St., 49
Mussolini, 72

Nectarios of Aegina, St., 61-65

Index of Proper Names

Nicodemos the Hagiorite, St., 49
Nietzsche, 70-72

Pasteur, Louis, 23
Perry, Ralph Barton, 12-13
Plato, 41, 48, 52, 58, 59, 60, 65-66, 83
Pope, Alexander, 57-58
Price, Weston A., 76, 77-79, 82

Rignano, Eugenio, 11

Saint-Hilaire, Geoffroy, 24
Santillana, Giorgio de, 10, 11
Seraphim of Sarov, St., 66
Sinnott, Edmund, 42, 52-53
Socrates, 41, 58, 59
Sounkras, S.P., 63
Symeon the New Theologian, St., 49

Vasileiadis, Nikolaos P., 70-72

Walker, Kenneth, 38-39
Wallace, Alfred Russell, 49-53
Weismann, August, 21
Whitehead, Alfred North, 73-74, 82-84

INDEX OF SUBJECTS

acquired characteristics, 16-21, 25, 27

aesthetic capacities of man, 50, 58, 61, 73

anthropology, 57-58, 65

   physical, 10, 57, 78, 80

ape-man (*pithicanthropos*), 56-57

apes, 45, 55-59, 61, 64

archetypal forms, 47-49

assumptions, 21, 23, 32, 34-35, 41-42, 44, 45, 47, 82

atheism, 22, 23, 70, 71

biology, 9, 10, 11, 12, 18, 76

botany, 9, 15, 54

Byzantine theologians, 48-49

chance, 40-44, 64

classification of species, 34, 45-46

Communism, 25-26, 71

comparative vertebrate anatomy, 9, 24, 54

consciousness, 52

cosmic plan, 22, 51

creation, 22, 25, 34, 46-49

Index of Subjects

creationism, 23, 28, 47-51

degeneration of man—mental, moral, physical, 75-82
deism, 23-24, 68-69
Design, 23, 28-29, 39
dinosaurs, 36
dogmatism, 30, 41, 76, 82, 84

elephants, 38-39
embryology, 54
environment, 16, 25, 26, 47
ethics and evolutionism, 68-70, 73-74, 83

Fascism, 72
fictions, 21, 35, 64
fossils, 46, 47, 54

genetics, 21, 43, 54
geology, 8, 47, 54
germ plasm, 21
giraffes, 17-20, 38-39
God, 11, 22-25, 27-29, 39, 41, 46-52, 59-60, 68-69, 71
gorilla, 45, 56, 57

Harvard, 8-13, 46, 57, 78, 82
heredity, 16-17, 20-21, 25, 27, 33-34, 43
hypothesis, 19, 35-36, 53
hypothetical mode of reasoning, 32-36, 53

imaginary illustrations of evolution, 17-20, 33
imagination, 19, 23, 27, 33, 45, 46, 54, 57, 82
inverse of evolution, 75-82

Java skull, 55-56
Jews, 72

life force, 41
life, origin of, 10, 11

man, 10, 11, 24, 29, 49, 55-66
mastodons, 36
materialistic explanations, 11, 28, 70, 73-74
mechanistic cosmology, 24
mechanistic explanations, 11, 28-29, 39-41, 44
mind, 39-41, 49-51, 59-60, 64
"missing links," 55-59
modern natural science, 83, 84

mollusks, 43-44

moral capacities of man, 50, 58, 60-61

morality, 68ff., 77

morphology, 54

"Mutual Aid," 37-39

"Natural Selection," or the "Survival of the Fittest," 30-34, 39-40, 42, 43, 50

Nazism, 71-72

Neanderthal skull, 56

Neo-Darwinism, 12, 28, 44

Neo-Lamarckism, 12, 25-26

new species, 17, 19, 21, 27, 29, 32-36, 54

order of Nature, 12

origin of life, 10, 11, 22, 24, 28

origin of man, 10, 11, 29, 55-56

origin of species, 15, 29, 32-34, 42, 45, 54, 64

orthogenesis, 44

paleontology, 47, 54

pectin, 43-44

Peking skull, 56

philosophical speculation, 17-18, 19, 22, 54, 82-83

philosophy, 10, 11, 12, 13-14, 35

physical anthropology, 9, 10, 57

postulates, 24, 44, 57

preconceptions, 26, 60-61

presuppositions of the natural sciences, 12

pithecanthropos, 56-57

purpose, 39-41, 44, 51, 53

rationality, 58, 60-62, 84

religion and evolutionism, 61, 68-74, 83

revelation, 65-67

Russia, Soviet, 25-26

scientific method, 10, 12

shorebirds, 17-20

"social progress," 25-26

soul, 50, 52-53, 58, 59, 62-66, 76

species, 29-34

speculative thinking, 18, 19, 22, 58-59

spiritual capacities, 58, 61-62

spontaneous generation, 22-23, 41

stability of the species, 24-25, 36-37, 46-47

Index of Subjects

"Struggle for Existence," 30-31, 37-38, 50, 71, 72, 74
"Survival of the Fittest," 30-31, 37, 71, 72

transmission of acquired characteristics, *see* heredity

use and disuse, law of, 16-19, 42-43

values, 73-74
"variations," 73-74
varieties, 29-30, 32, 34, 36-37, 42-44, 54, 75
"vital force," 25

Wheaton College (Mass.), v, 14

Zoology, 9, 15, 46, 54

OTHER PHILOSOPHICAL BOOKS
BY CONSTANTINE CAVARNOS

A DIALOGUE BETWEEN
BERGSON, ARISTOTLE, AND PHILOLOGOS

A comparative and critical study in the epistemology and metaphysics of Bergson and Aristotle. Preface by Professor John Wild. First Bowdoin Prize (Harvard University) among the graduate dissertations of 1947. Cambridge, Massachusetts, 1949. 3rd, enlarged edition, Belmont, MA, 1988. "Of all literary forms, that of the dialogue is perhaps best fitted to aid and intensify the philosophic quest. Dr. Cavarnos in this Bowdoin Prize Essay has used this form in a masterly way to raise certain questions and to examine certain doctrines which are of great contemporary interest... What is that ever present dynamism and change with which the whole world of nature is always pulsating? What is that rational insight or awareness which is the peculiar possession of man? These basic questions set the dominant themes for this dialogue. They are developed with a lively style and wit which never lets the reader down. Dr. Cavarnos not only knows how to write: he also has something to say. In addition to many illuminating comments on Bergson, one of the most influential thinkers of our time, he has also succeeded in revealing the amazing relevance of classical Greek thought to present-day problems. This dialogue deserves a wide audience. Every reader is likely to become a participant."
—John Wild, *Harvard University.*

"A university life reacts on an intellect according to the man's particular substance. The routine mind reproduces a stereotyped, dull thesis that promptly becomes a dust collector, while the keen, original mind brings forth something new and vital that sets off a continuous wave of reaction on other minds. In the latter category is *A Dialogue between Bergson, Aristotle, and Philologos* by Constantine Cavarnos. —*Athene.*

"This essay is witty, lively, and acute."
—Raphael Demos, *Harvard University.*

MAN AND THE UNIVERSE IN AMERICAN PHILOSOPHY

This book deals with the problem of what man is and what he can become, and how he is related to the rest of reality, as this problem is confronted by three great representative American philosophers: Emerson, James, and Whitehead It is based on four lectures which Mr. Cavarnos delivered in Greece during his tenure as a Fulbright Research Professor at the University of Athens (1957-1959). Athens: "Astir" Publishing Co., 1959.

"*Man and the Universe in American Philosophy* is written in lucid modern Greek... It is an excellent introductory account of three philosophic giants of American thought—Emerson, James, and Whitehead—and a handy analysis of recent trends in American philosophy."
—John E. Rexine, *Colgate University.*

"One must be very grateful to C. Cavarnos for the information he gives and for the concise, precise, clear, and well-written exposition of philosophical ideas in America, which this book presents."
—Bambis D. Klaras, *Spoudai ("Studies")*

"This book must be recognized as the first comprehensive monograph of its kind to make available to the Greek-speaking public some of the representative thinkers and ideas for which American thought has won renown. It is a well-written book, and Dr. Cavarnos is to be commended not only for the merits of his expository abilities and appreciative insights, but also for his effort to translate judiciously his own direct knowledge of our intellectual heritage.

...The reader will unhesitatingly recognize the merits of this book's admirable account of Whitehead's views as well as the sympathetic understanding of both Emerson and James. —John P. Anton, *Emory University*.

MODERN GREEK PHILOSOPHERS ON THE HUMAN SOUL

Selections from the writings of seven representative thinkers of modern Greece on the nature and immortality of the soul, compiled, translated and edited with a Preface, Biographical Sketches and Notes by C. Cavarnos. The philosophers presented are Benjamin of Lesvos, Vrailas-Armenis, Skaltsounis, St. Nectarios, N. Louvaris, Ph. Kontoglou, and Theodorakopoulos. Belmont, MA, 1967. 2nd, considerably enlarged edition, 1987.

"A significant book... It calls attention to important recent and contemporary work in an area which is usually supposed to have existed only in ancient times."
—George B. Burch, *Tufts University*.
"This is a very beautiful publication. The texts have been very well translated. And the introduction is extremely good."
—I. N. Theodorakopoulos, *University of Athens*.